15.96

WITH

DATE DUE

APR 2 4 1998

CANYONS

RANDY FRAHM

CREATIVE EDUCATION

Designed by Rita Marshall
with the help of Melinda Belter

Published by Creative Education,
123 South Broad Street, Mankato,
Minnesota 56001.

Creative Education is an imprint of
The Creative Company

Photography by Carr Clifton, F-Stock,
Gerry Ellis, Minden Pictures, Peter
Arnold, Inc., and Photo Researchers

Library of Congress
Cataloging-in-Publication Data

Frahm, Randy.
Canyons / by Randy Frahm.

    p.   cm.
ISBN 0-88682-707-8.

1. Canyons—Juvenile literature.
[1. Canyons.]   I. Title.   94-3156
GB562.F73   1997   CIP
551.4'42—dc20   AC

5 4 3 2 1

Printed in Hong Kong

7

When a river fights rock, the earth cracks under the pressure of the battle. These cracks do not become scars, but beauty marks, places where water runs with the sound of thunder and rock rises into the sky higher than the tallest building. These cracks are called *Canyons*, and their size, depth, and majesty place them among the earth's most awe-inspiring natural features.

Canyonlands in Utah.

The word "canyon" comes from the Spanish word *cañon*, meaning "tube." A canyon is a type of river valley called a *Gorge*, meaning that it has a relatively narrow top and a sharp "V" formation. Although people often call any small canyon a gorge, scientists make a distinction between the two. Canyons have a steplike appearance because the rock erodes at different rates. Non-canyon gorges, on the other hand, have a flat profile because the rock erodes at a similar pace. Thus, while all canyons are gorges, not all gorges are canyons.

The stepped walls of Deshutes River, Oregon.

To fully understand how a canyon or gorge is formed, we must first understand how rivers create valleys. As a river runs from its start, called its *Source*, to its end, or *Mouth*, it erodes a channel through the land and eventually forms a valley. The shape of a valley's walls are determined by the type of rock in the river's bottom and sides. Soft rock erodes faster than hard rock. If the sides of the riverbed are soft, they will wear away quickly and evenly and the river channel becomes wider. Periodic flooding of the river may also widen the bottom of the valley and reduce the steepness of its sides. The result is a broad, U-shaped valley. Hard rock,

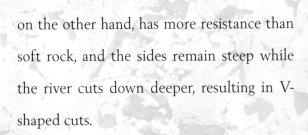

"

on the other hand, has more resistance than soft rock, and the sides remain steep while the river cuts down deeper, resulting in V-shaped cuts.

—

Another factor in the formation of a valley is the speed, or *Velocity*, of the river. Slow-moving rivers carve gentle, broad valleys. Fast-moving water cuts down deep into the land beneath, gouging out steep, narrow gorges. The deepest valley in the world measures 3.75 miles (6 km) from top to bottom. It was formed by the Kali Gandak River in Nepal, among the Himalayan mountains.

Malad River Gorge, Idaho.

Many other geological processes are involved in the formation of canyons and gorges, and several of them can combine to create these marvels of nature. *Uplift* is one of the most common contributing factors. The earth's crust is divided into sections called *Plates* that move at an extremely slow but constant pace. Uplift occurs when two of these plates squeeze together and rise upward. The Colorado Plateau—which covers the northern half of Arizona, northwestern New Mexico, southwestern Colorado, and the southeastern half of Utah—was uplifted when the Rocky Mountains first formed.

When uplifting occurs under an existing riverbed, the river's slope becomes steeper and the river becomes faster. It now has more power to cut downward. This process, called *Rejuvenation*, affected the Colorado River when the Colorado Plateau was uplifted. In the last 120 million years, the Colorado River has cut as deep as 5,940 feet (1,810 m) into the Colorado Plateau, forming the Grand Canyon, Glen Canyon, Marble Canyon, and others. The deep gorges formed by various rivers in West Virginia and Pennsylvania are also the result of uplift, this time of the Allegheny Plateau.

A river winds through the canyonlands, Utah.

Some canyons and gorges appear to defy the laws of nature. Rivers, and thus canyons, usually follow the path of least resistance, following along the bottoms of the natural faults in the earth's crust. But *Transverse Canyons* appear to do the opposite, as if the rivers that made them decided to take the most difficult route. Transverse canyons are found all over the world, from the Andes mountains of South America to the Zagros Mountains of Iran.

Uplift plays an important role in the development of transverse canyons. As land rises under an existing river valley, new mountains, ridges, canyons, and gorges develop around the river channel and eventually produce a river that appears to have chosen a difficult path. Uplift can also work with *Superimposition.* In this process, *Sediment,* or particles of rock and organic debris, covers existing land forms. When a river starts, it slices into the sediment and creates a new valley. The sediment covering the old land forms can be eroded away over time, exposing them again and making the younger river valley appear as though it formed against the natural drainage patterns of the surrounding landscape.

Strange rock formations in Bryce Canyon, Utah.

16

Stream Piracy is another process that affects the development of canyons and gorges. It occurs when one river erodes through its valley into the valley of a second river. Often the two rivers combine to become one, and the two valleys endure major changes. The valley that has been robbed of its water now appears inexplicably large and steep, since its river is gone. The valley containing the now much larger river will gradually expand and become deeper to accommodate its increased water, and new gorges and canyons may be formed. Many of the gorges in the Appalachian Mountains of the eastern United States were created by stream piracy.

The Escalante River, Utah.

18

Climatic changes can also contribute to the carving of canyons and gorges. Nearly two million years ago, a gradual cooling of the earth's atmosphere led to the formation of *Glaciers*—sheets of snow, ice, and earthen debris so heavy that they slowly moved from the force of their own weight. Glaciers covered much of the planet and became a major force in creating our landscape. They even changed the river courses, and thus affected the development of canyons. For example, the Little Missouri River originally flowed north through North Dakota to Hudson Bay in Canada. As glaciers approached from the north, they pushed the river's course east and increased its power. As a result, the river helped carve out the Little Missouri Trench in North Dakota, an area with gorges, canyons, and steep hills also known as the North Dakota Badlands.

North Dakota Badlands.

Among the most spectacular features of canyons are violent waters known as *Rapids.* A riverbed usually narrows when it enters a gorge or canyon, even though it carries the same amount of water. The result is similar to what happens when a person places a thumb over the end of a garden hose—the water flows much faster. This furiously churning water can be found in many canyons of the United States. The New River Gorge in West Virginia has the highest concentration of rapids, 21 in a length of 15 miles (24 km). In rapids, the water also tends to force its way downward, and the river itself can become deeper. The Yangtze River in China is known for its deep gorges and rapids. In some areas, the depth of the river reaches between 500 and 600 feet (152 to 183 m), making the Yangtze the deepest river in the world.

Wuxia Gorge, Yangtze River.

While water running through valleys travels fast, the rocks of canyons and gorges hold time beautifully still. Consider, for example, the many types of rock that make up Arizona's Grand Canyon. The oldest type of rock is vishnu schist, which dates back 1.7 billion years and is often polished shiny by wind and sand. A graceful, swirling type of rock named dox sandstone was once part of a tidal area of an ocean that covered the region more than 330 million years ago. In fact, fossils of early sharks and coral, common in today's oceans, have been found in the layers of limestone in the canyon.

Eroded sandstone formations.

24

As a canyon or a gorge drops from its highest point, the changes in elevation affect the weather and the plants and animals that live there. In this way, a canyon can hold life forms representative of many different areas. Descend 6,000 feet (1,830 m) from the north rim of the Grand Canyon and life forms usually found from central Canada to central Mexico will be encountered. Near the top of the canyon, the temperature and climate is similar to cooler Canada, and ponderosa pines and Douglas firs are found. But as the elevation drops, aspen trees replace the fir trees, only to give way eventually to the desert cacti that live near the river on the canyon floor. Animal life can range from mule deer and mountain lions in upper elevations to desert reptiles such as bull snakes and horned toads on the canyon floor.

Page 24: Winter in Bryce Canyon, Utah.
Page 25: The prickly pear cactus, canyon vegetation.

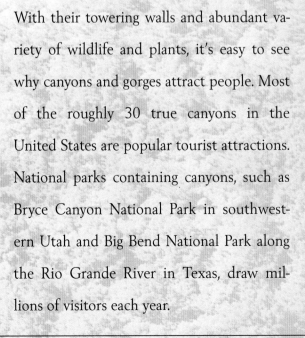

27

With their towering walls and abundant variety of wildlife and plants, it's easy to see why canyons and gorges attract people. Most of the roughly 30 true canyons in the United States are popular tourist attractions. National parks containing canyons, such as Bryce Canyon National Park in southwestern Utah and Big Bend National Park along the Rio Grande River in Texas, draw millions of visitors each year.

Genesee River Gorge, New York.

Perhaps the most popular canyon in the world is the Grand Canyon. Nearly three million visitors arrive at the Grand Canyon National Park each year to view this amazing tribute to the power of the Colorado River. Nearly 280 miles (450 km) long, as much as 18 miles (29 km) across and reaching down over a mile (1.6 km) into the earth, the Grand Canyon is considered one of the most spectacular sights in the world.

Ironically, however, the popularity of this natural wonder actually threatens it. Visitors often leave litter on the floor of the canyon along the banks of the Colorado River or remove rocks or plant life, which can affect the look of the landscape as well as contribute to erosion. To help protect this and other canyons, legislation now makes these activities a serious offense, and the number of rafters and campers has become more tightly regulated.

The Grand Canyon, Arizona.

31

As we strive to protect and preserve the vast resources of our planet, we must be sure to include canyon environments in our efforts. *Canyons* reveal to us steep cliffs, raging rivers, long-dead oceans, and geological history. They are testimony to the creative forces of rivers, the strength of lifting earth, and the power of glaciers. Within their towering walls, canyons whisper the secrets of our ever-changing planet. If we listen carefully, perhaps we can hear them.

Snake River at Hells Canyon, Idaho.

Index